Robert Stewart Norris

On Paranitroorthotolylphenylsulphone

and some of its derivatives

Robert Stewart Norris

On Paranitroorthotolylphenylsulphone
and some of its derivatives

ISBN/EAN: 9783337339647

Printed in Europe, USA, Canada, Australia, Japan

Cover: Foto ©berggeist007 / pixelio.de

More available books at **www.hansebooks.com**

On

Paranitroorthotoluyl ....
sulphone and some of
its Derivatives.

A
Dissertation
Submitted to the Board of
University Studies of the
Johns Hopkins University
for the Degree of Doctor
of Philosophy.

By

..... Stewart .....

1877

Contents:

Acknowledgment.

Introduction.

Method.

Preparation of Material
    Paranitroorthotoluen...
    s.. p....   ....a.

Paranitroorthotoluene
    sulphonic bromide.

Paranitroorthotoluidien
    ..lphone.

Preparation and Properties.

Action of fuming sulphuric
    Acid.

Paranitroorthotoluy-
    phenylsulphone-
    sulphonic Acid.

Action of Potassium Hydroxide.

.tion of ..   ...

ammonium ser...ide
Paramidoo thotoly...
  ...my sulphone.
...tion of Oxidising ...gents.
Chromic ...d.
Potassium ...manganate.
...ranitroo trophenyl subhone-
benzoic Acid.
Salts.
  Barium salt.
  Calcium salt.
Comparison with P...a
nitroo thotony...y....
sulphon....,
...tion of Phosp...
...imer...d.
P...nitroo tho...my
sulphon...muoyl Oxid...e.
...tion of Water...

Acknowledgment.

This investigation was
undertaken at the suggestion
of Professor Ira Remsen, and
carried out under his
supervision. The author
wishes to thank him for the
advice and assistance
received during the prosecu-
tion of this and previous
work, and also for instruc-
tion received in the lecture
room, both of which have
been a constant source
of inspiration.

On Paranitroorthosulphbenzyl-
sulphone and Some of its
Derivatives.

Introduction

In a series of investigations
carried on in this laboratory
on derivatives of orthosulphine
benzoic acid, it has been
shown that when its
potassium salt is treated
with phosphorus penta
chloride two isomeric
chlorides are formed[1], the
formulas of which have
been determined to be,

$$C_6H_4 \underset{SO_2}{\overset{COCl}{\diagdown}} Cl \text{ and } C_6H_4 \underset{SO_2}{\overset{COCl}{\diagdown}} O.$$

In treating either of these
chlorides with aluminium
chloride and benzene, one

the chlorine atoms is replaced by a phenyl group; and the products formed in both cases are identical[T]. There must, therefore, be a rearrangement of the grouping of the atoms in one of the reactions.

There are three possible formulas for the product,

$$C_6H_4 < \frac{COCl}{SO_2C_6H_5}, \quad C_6H_4 < \frac{COC_6H_5}{SO_2Cl}, \quad C_6H_4 < \frac{?}{?}$$

     I.   II.   III.

From its properties and reactions the most probable structure seems to be that represented by the second formula, the phenyl residue being in combination with a carbonyl group here as

[T] Remsen and

and Saunders' converted the chloride into the corresponding sulphonic acid or replacing the chlorine with a hydroxyl group; prepared a number of derivatives from these; and studied their properties. They also obtained a diphenyl derivative of orthosulphobenzoic acid, of the formula,

$$C_6H_4 < \begin{matrix} CO C_6H_5 \\ SO_2 C_6H_5 \end{matrix}, \text{ by the more}$$

prolonged action of benzine and aluminium chloride on the dichloride. The argument which they bring forward in favor of the structure presented by the formula

$$C_6H_4 < \begin{matrix} CO C_6H_5 \\ SO_2 \end{matrix} \text{ for the ...}$$

a ohect of the reaction."

1. That the acid formed by the decomposition of the chloride with water is a very soluble substance, and difficult to obtain in a crystalline form.

This is ~~very~~ characteristic of sulphonic acids in general, while, on the other hand, the corresponding carboxyl acids are much less soluble, and easily obtainable in the form of crystals. Paraphenylsulphone benzoic acid, for instance, $C_6H_4 \begin{smallmatrix} COOH \\ SO_2C_6H_5 \end{smallmatrix} (p)$ described by Hewell, is a difficultly soluble, crystalline compound; and another example will be described later in this chapter.

2. That the decomposition of
the potassium salt by fusion
with caustic potash gives
benzoic acid and benzene
sulphonic acid.

It is true, that on purely
theoretical grounds this
might be a possible reaction
with either of the substances
represented by the formulas,

$$C_2 H_4 \begin{matrix} CO_6 H_5 \\ SO_2 OK \end{matrix} \quad \text{and} \quad C_2 H_4 \begin{matrix} CO_2 K \\ SO_2 C_6 H_5 \end{matrix}$$

$$I. \qquad\qquad\qquad II.$$

But when we come to consider
the experimental evidence on
the point it seems much more
probable that these products
would result from a compound
having the first formula, than
one having the second, since a

sulphone, for instance, when
fused with caustic potash
breaks down completely into
diphenyl, sulphur dioxide, and
phenol[1], and various other
substances. Polysulphone acts
in a similar manner[2]. And
these reactions take place with
difficulty, showing the stability
of the sulphone grouping. —

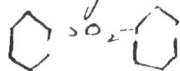

The way in which the diphenyl
substitution-product of ortho
sulphobenzoic acid breaks down
also furnishes evidence in
favor of this view. They
found that it decomposes
readily under similar
circumstances into two

acid and diphenyl or ston,

$$C_6H_4 \begin{cases} COC_6H_5 \\ SO_2C_6H_5 \end{cases} + H_2O = C_6H_5 \cdot C_2 \cdot C_6H_5 + C_6H_4 \cdots$$

3. That the ethyl ether breaks
down very readily on
standing in contact with
alcohol, into the acid; which
points to its being an ester
of a sulphonic acid rather
than of a carbonic acid.
4. That the chloride on treatment
with ammonia, instead of
forming an amide, goes into
the lactim condition. It is
in accordance with the
supposition that the carbon
atom left in the molecule is the
one united with the carboxyl
group.

The evidence, therefore, is as

pretty clear, that of the two formulas $C_6H_4\diagdown\overset{o\;C_6H_5}{SO_2Cl}$ and $C_6H_4\diagdown\overset{SO_2Cl}{SO_2C_6H_5}$, the former is the more probable for the compound obtained from the isomeric chlorides of orthosulphobenzoic acid by the Friedel and Crafts reaction with benzene.

Paranitroorthosulphobenzoic acid has also been found to yield two isomeric chlorides on treatment with phosphorus pentachloride,

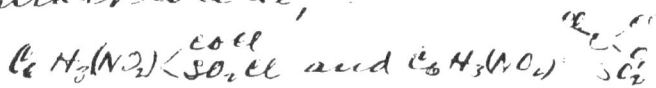

$$C_6H_3(NO_2)\diagdown\overset{COCl}{SO_2Cl} \text{ and } C_6H_3(NO_2)\diagdown\overset{CO}{SO_2}$$

These in turn yield two new phenyl substitution products, when treated with benzene and aluminium chloride,[2] giving the three possible

ormulas,

$$C_6H_3(NO_2) \overset{CO \cdot C_6H_5}{\underset{SO_2Cl}{}}, \quad C_6H_3(NO_2) \overset{Cl}{\underset{SC_2C_6H_5}{}},$$

I II

$$C_6H_3(NO_2) - SO_2 \overset{Cl}{\underset{O}{\diagup C_6H_5}}$$

III

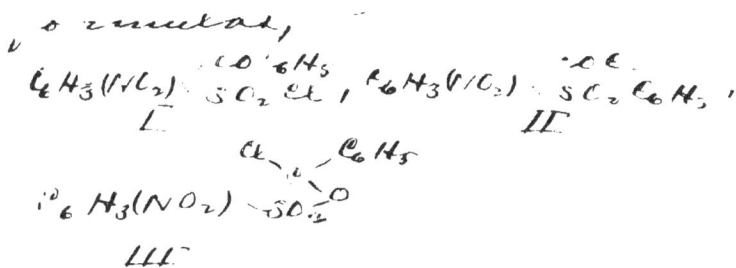

But the reaction seems in this
case to stop here, no diphenyl
substitution-product is formed
by the further action of the
reagents.

The same question arises
in this case as in that of the
compound just discussed.
Which of the chlorine atoms
is replaced by the phenyl group?
Hollis[@] investigated the product
formed, and found from its
properties and reactions
some what similar experiments

in favor of the first formula,
as that found by Remsen and
Saunders in the former case.
The acid obtained by decomposing
the chloride with water or acids
was very soluble, and could not
be obtained in the form of soda.
The ethyl ether of the acid was found
to be so unstable, if one exists
at all, as to pass over immediately
into the acid, when an attempt
was made to prepare it from
the chloride and absolute alcohol.
In treating the chloride with
ammonia a reaction was found.
These three reactions are perfectly
analogous to those which take
place with orthobenzoylbenzene sulphon
chloride under similar conditions.

and point to the structure

for the chloride. But the most
conclusive evidence obtained by
Remsen and Saunders con t
formula which they gave, was
the products formed by the decom-
sition of the chloride on fusing
with caustic potash, was not
duplicated in the present case.
The presence of the nitrogroups
would probably make such a
test less satisfactory. The view
in this case is not as convincing
as it is in that of the more
simple compound.

There is, however, another
method of attacking the problem,
by which just as conclusive
evidence, but of a different

kind might be obtained. It was
for the purpose of determining
the nature of this evidence that
the present investigation was
undertaken.

If a compound with the
formula [structure: benzene ring with COOH / SO₂NH₅ and NO₂] could be prepared,
and its properties should be
different from that obtained by
Hollis from the isomeric
chlorides of paranitroorthosulpho
benzoic acid, the latter could no
have the same structure, and
therefore would probably have th
formula [structure: benzene ring with COOH / SO₂Cl and NO₂]. If their properties
should be the same, the struct
of the two would, of course,
identical.

The best method wi

suggested itself for preparing

paranitroorthophenylsulphone-

carbonyl chloride was to start

with paranitrotoluene,

convert it into paranitrotoluene-

orthosulphonic acid by treatment

with fuming sulphuric acid,

t. at the potassium salt of it

with phosphorus pentachloride

to form paranitrotoluene the

sulphone chloride, from which

paranitrotolueneorthophenyl

sulphone could be made by the

Friedel and Crafts reaction

with benzene, by oxidation of the

methyl group, paranitro the

phenyl sulphone benzoic acid

would be formed, and this

could be transformed into

... and introduce new ... mon ...
... experiments, by the action
of ... phosphorous ...

... becomes necessity to use
a long series of transformation;
in case they all look ... , ...
an abundance of material
to start with, in order that
the end product might be
obtained in sufficient quantity
for investigation ...
... obtained at ...
stages be small.

Another method was tried,
starting from ...
... 40
... was to be converted into
the ...) ...
the ...

...paration of ...t...ial,

...r anitrolui... eo...hose a.e...'

...id.

This acid was prepared by th

method originally describ... ...

...eistein and Kuhling[1], and ...t.

used by Hart, Hartl, ...ay ...

others[2] in this ....oratory. Th...

are a number of importan...

details which are not to ....d

mentioned in any one artic...

and which it is ....

to observe in order to obtain

... good yield, and a ... ...

...roduct, so it may b... not ...

out of place to give the ...tai...

of the process by which th...

...ults .... attained.

Fourteen .... grams ...

pure ...
... in a ... it ...
... hundred grams of ...
... sphuric acid added. The ...
... phuric acid contained in the
origin ... ... 's
... sufficiently strong for the ... ,
but acid which has been standing
for any length of time in the
... if ... they ... on ...
... ...
... so much ... as
to be unfit for ... ...
... ration. The ... of
... phuric acid to the ...
... for ... is to ...
... a did in the ... ...

The flask was then placed
in a large water-bath so as to

...immersed in th... ...
to the mark, and th... ...
...d to boil a ... ...
of fifteen to twenty minutes
the solution was thorough...
agitated by shaking th... ...,
and after the solution had ...
treated for two or three now
a sample was occasionally
removed from the ... and
added to water to determine
when the ...tion was complete,
which was indicated by its
...ing entirely dissolved, ...
time taken to comp... ...
...phurize the nitro... ...
...e was ... on th...
strength of th... ...
... ... too ... ...

At the end of this time. The
solution was allowed to a
down to the ordinary
temperature, the dark
syrupy liquid obtained in
this manner was poured
into a vessel with an the
water contained in any
evaporating dishes, the
were not more than
thirds of the solution
has taken, in order
the thoroughly mixed
solution should be quite
even and free from
undissolved matter. if in
this not the case, the
solution should be if
in circum was consta

... is no ... to
... solution until no
further effervescence too...
... are, the solution being
stirred during the ...
The precipitated calcium
... and excess of
calcium carbonate were
then filtered off through
a cloth ... This did no
... of the ...
so it was necessary ...
again through ... , in order
to prevent the formation ...
a large amount ... potassium sulphate in the
... stage of the process, ...
... ...

... crystals ... ...
with water, and the ... as
... liquid ... off ... ...
a filter, and added to the
main solution, when ...
then ... to boiling, a
solution of potassium carbonate
was added until litmus ... ...
... a slight alkaline
reaction, and the precipitated
calcium carbonate filtered
off. The dark brown solution
was boiled for an hour with
animal charcoal, and again
filtered. The removal of the
animal charcoal was found
to be a slow process, and
it was accompanied ... ...
... a ... ...

The use of ... ...
with a ... filter ... was ...
... the process, ... Will
... mate covered with a ...
filter paper ... as d... in con... ...
with a filter pump, and
... vious filtration thro... ...
... oth were tried, but to
print d filter paper was foun...
to be most expeditious. ...e
... o... did ... ot ...n ow
... of the co... from th...
solution, which was st...
brownish yellow. ... wa...
...vaporated to o... ha... of ...
o... ig... al volume, ... th...
potassium salt a... owed t...
crystallize out o... ... ...al ...
th... ...lution; it... ...d ft... ...

... first coat ...
... other one ... ed, ...
... potassium ... so ... large ... ed
out, and so on, until ...
volume of the solution
... equal to ...
... when it was again
treated with animal charcoal,
filtered and evaporated so
to one litre. ... not ...
... after the crystals had
separated out of this solution
... thrown away, ...
crystals formed ...
... water, the crystals
which separated out at
different times, were ... ed
as thoroughly as ...
... filtration, ...

or a q s ... ej ... by ... the ... ... and ... ... dry. In this condition ... appeared as very ... ... Sixteen and ... twenty two grams of ... salt were obtained from the four hundred grams of paranitrotoluene ... d.

It has been shown up... [*] that the compound form... be the action of fuming su... aci... aid on paranitrotoluene is para nitro o thoto... ... su... oxic acid. ... converi... the acid form ... on in's ...action, $C_6H_3 \begin{smallmatrix} (H_3 \\ (C_3H \\ (C_2 \end{smallmatrix}$, into amido

[*] Ann ... ...

... ... $C_6H_4 \begin{cases} C_6H_3 \\ H \end{cases}$
... ... tion of the nitro group
which is known to ... in the
... a ... position with ...
t. the methyl group ... on
account of the formation
of the acid from ... a nitro
toluene. This was diazotised
to paradiazotoluenesulphonic
acid $C_6H_3 \begin{cases} CH_3 \\ SO_3 \\ N \\ N \end{cases}$, which was
converted into toluenesulphonic
acid by boiling with absolute
alcohol, according to the
following reaction,

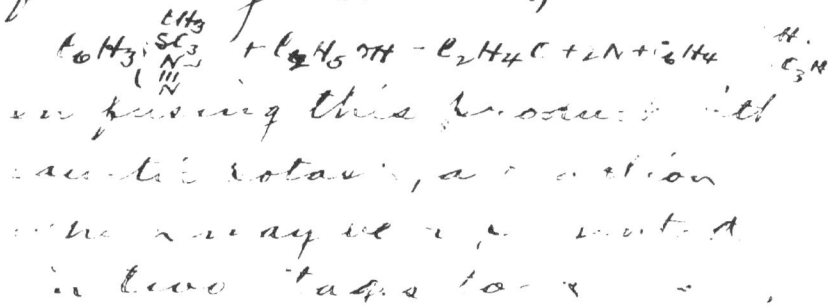

$$C_6H_3 \begin{cases} CH_3 \\ SO_3 \\ N \\ N \end{cases} + C_2H_5 OH = C_2H_4O + 2N + C_6H_4 \begin{cases} H \\ C_3H \end{cases}$$

on fusing this ... ...
... rotary, ...
... ...
... two stages to ... ,

$C_6H_4 \diagdown \begin{smallmatrix} H \\ C_6H \end{smallmatrix} + KN + HK : + C_6H_4 \begin{smallmatrix} H \\ CH \end{smallmatrix}$

$CH_1 \diagdown \begin{smallmatrix} H \\ OH \end{smallmatrix} + KOH + H_2C \diagdown C_6H_4 \diagdown \begin{smallmatrix} OOK \\ OH \end{smallmatrix}$ etc.

and the product so ... was found to be the potassium salt of ... acid, $C_6H_4 \diagdown OH$ therefore the nitrotoluen-sulphonic acid for ... a from ... raw had the structural formula,

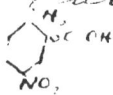

and the compound obtained ... ... above described the potassium salt of the ... dinitolune ... ... , ...

paranitrotoluene ...

... Jensen[1] ... ated the ...
... of paranitrotoluene ...
... phonic acid to 110°C, and
found that it ... no ...
in weight. It is ...
... necessary to ... at ...
... before ... ing ...
... The method made use of in
preparing the chloride was
also first used by Jensen.[2]
Four hundred and sixty grams
of the potassium salt of
paranitrotoluene ...
acid were ...
... portions ...
... ...

... mix ... in a mortar ... ...

... up, and reaction began when the temperature reached ... 100° ..., continuing at the same temperature, ... The ... ... temperature fo ... half an hour, in order to ... off most of the ... . allowing it to cool to the ordinary temperature, it was poured into ... ... containing ... ...

... on a
metal disk on ... ...
gauze ... od. The warm ...
... the chloride in a ...
state, and it can be ...
thoroughly washed in ...
condition than of solid.
After allowing the ...
to settle the wash-water was
decanted or siphoned off,
more warm water added
and the process of washing
repeated. Cold water was
finally added, ...
it ... up with ...
chloride immediately ...
It was then ground up into
a fine powder in a mortar,
and was ... with ...

... ...

... ...

... ... was ... ...

phosphoric acid. The ...

... ... off ... ...

of a filter-pump, ...

... oxide dried with ...

paper. 370 grams were obtained

The ... ...

sulphon chloride prepared

in this way was ... ...

powder, melting ... ...

and readily solu... in ...

... which ... ...

... large, rhombic crystals

melting at 44° C. uncor. ...

a number of different ...

of manipulation ... ...

... ... to ...

good results at too ... too
high a temperature, or ... too
too long increases the yield, and
makes the chloride dark. The
purity of the product was
a great deal over that of the
... was ... the
... of the latter was found
to be carried over into the
chloride.

... chloride ...
... chloride ...
... the ...
... boiling-point of the
product prepared by the
above process were
crystallised from ...
... ...

...th ... ... in ... ...
state ... the ordinary
temperature ...
... to some ...
... it ... is
... ...
working with ... ...
... that th... ... with
... any particular ...
... ... to crystalli...
... much ... ...
... ...

4 amidoo ...

$$C_6H_3 \cdot \overset{CH_3}{C_2H_3}( )$$
$$(NO_2(4)$$

This was obtained from
paranitro... toluene
... chlorine ... by means
of the Friedel and Crafts
action with benzene.
Limpricht[C] obtained a
... compound isomeric ... ...
this, by the action of
aluminium chloride on
toluene and meta-nitro ...
sulphone chloride,

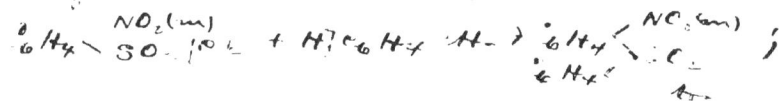

$$C_6H_4 \overset{NO_2(m)}{<} SO_2 ... + H|C_6H_4 ... H \rightarrow C_6H_4 \overset{NO_2(m)}{<} ... ;$$

properties which ... ...,
it appears to be quite
different from ... nitro...
...oxyben...; such are,
kept in solubility, ...
these points of differ...
... be taken ... ...
the subject of the ... ... of
oxidising agents on it
... .

<u>Preparation.</u>

As a preliminary exp...
... in grams of ...
... ...
... 100 c.c. of ... ...
a small flask ... ...
with a condensing-tube,
... four grams of ...
... ...

The residue left aft... evaporating
off the benzene... a...
brown, tarry mass...
treated in different... times
with alcohol, ether, chloroform
and benzene, and th...
solutions rendered, crystals
separated from... of...
cautious on... evaporation,
but the... and...
dissolved mass of the...
tarry substance,...
consequently,... not...
not regular;

It was found on...
retain it... that the
aluminium chloride...
more satisfactory...
...

@ Inaug. Diss. Johns Hopkins University!

with the ... ...

... ...

twohundred ... of ...

... ...,

from which ... ...

twohundred and ... ...

... of the ... ...

... ...

... ...

..., but the ... ...

... ...

the p... of ...

The collected residues were

... acted with hot ...

and the ...

... with ...

... ...

... ...

.

Analysis of the product
gave the following results,

I. 0.3140 grams gave 0.2352 grams _a

II. 0.1685 grams gave 0.1336 grams _a

III. 0.2865 grams gave 0.5734 grams _(

and 0.1658 grams +_

IV. 0.3444 grams gave 0.9201 grams _(

and 0.1008 grams H

V. 0.3229 grams gave 0.017 grams N.

	Calculated for $C_{13}H_{1}O_4N_5$	Found I.	II.	III.	IV.	V.
C	5ა.3C	---	---	56.5ა	56.0ა	-
H	4.00	---	---	4.13	ა.1ა	
ა	11.5ა	11.60	11.16	---	-	-
N	5.06	---	---	---	---	ა.ა.ა

Paranitroorthotolyl p-nitro-sulphone exhibits in a no it a degree the general stability of sulphones toward concentrated acids. It is, however, much less stable toward alkalies than the simpler sulphones, probably on account of the presence of the nitro group. Most of the ordinary organic solvents dissolve it

It was found to b
nearly as soluble in methyl
alcohol as in ethyl alcohol.
Benzene and acetone dissol..
it very readily, and it
deposited from the former
in the form of plates and
from the latter as needles.
It was almost equally
soluble in hot and cold ....,
and to a rather small exte...
hot ligroin dissolved it with
...., in water it is just a..
insoluble, even when the ..
is ....; but .... dissolves
dissolve to a ~~very~~ limited
extent is ... by the ...
that the fibres ... on ... ...
...

... ...

... small ... of the ... ...
dissolved in ... ...
... color, with
caustic potash solution.

... ...
sulphuric ... ...
sulphur in small amounts,
dilute hydrochloric ... did ...
not dissolve it. ... ...
nitric and sulphuric ... ...
... found to dissolve ...
... quite ...
... in the ... ...
... up again ...
with water. ... ...
was boiled with ... ...
... ...

was no appa... ra...n,
the ...te precipitate
...ed... dir...tion ...d ...
...tt...pl...

The sulphove disso...
...ite readily in ordina...
concentrated s...p...
in th...'d, ...nd ...
precipitated ...ci... yd...
th...point ... was p...
...to ...ate... th...n the...
was ...ed to 100...
...ct...t...ty of th... ...ph...
...ased, h...t ... ...a...
...ted up on. Wh... ...d
temp...at... ... ...
...... h...h...
... d...compos... ... ...
b... th... ... ...

A small amount of it

and ... to ...
... ...
... ...
... ... to the ... to
... ... as d to ...
... ... During the
digestion of the solution
... ... the solution a
...ate brown, and it increases
in intensity during the
...ting. This ...
again in the ... sol., ...
it was ... to ...
... the solution ...
It is the solution was allowed
to cool, and the various ...
crystallized ... as ...
... of the various ,
... the following ...

c.c. ... 4 grams ... ... grams ... ...

c.c. ... grams of t... ... ...

... c.c. ... grams ... ...

Calculated for	Found
$(C_{13}H_{10}O_7S_2N)_2 Ba + 6H_2O$	
H_2O 11.28	11.42
$(C_{13}H_{10}O_7S_2N)_2 Ba$	
Ba 16.16	16.35

<u>Paranitroorthotolylphenyl sulp... ... sulphonic acid.</u>

The acid was obtained ... ...
the barium salt by ... ...
a solution of t... ...atter with
just sufficient ... ...'d
to precipitate the bari... ...
barium su...phate, ...
... was found to ... readily
so... ble in water, ...

ized from the solution in the
form of very fine white needles.
They were readily soluble in
alcohol, but difficultly in
ether and benzene.

From analogy to the
sulphonic acid formed from
p-nitrosulphone, and from the
fact that three of the positions
on one of the benzene rings
are already substituted, it
seems probable that the
sulphonic acid residue enters
the phenyl group of the series
in the meta position, forming
a compound of the formula,

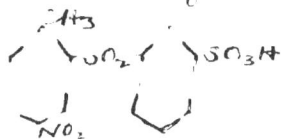

$$\underset{NO_2}{\overset{NH_3}{\diagdown}} NO_2 \diagdown SO_3H$$

...letion of potass... ...t...

on ... it ...

...

...ture of parasite...
orthotoluylphenylsulphone in
solution gives a ...ish
a color with caustic...

A few crystals of the ...
...dissolved in...
...a solution of caustic
potash ...ved. Upon ...
...that the solution qui...
assumed a deep purple color,
which became more...
after ...ing to...,
...and ...precipitate formed. ...
hydroxide...

...
...if... a portion of the
solution, on...
...or to reddish yellow, ...
...changed to a... bright
purple on... addition of... ...
of potassium hypochlorite. The
...ble colour... are not...
back again on ...

...the ...
sodium... carbonate to the
...id solution. Concentrated
sulphuric, nitric, and
hydrochloric acids ...
...produced a... ...
...
...
...

... ... did ... ...
The ... ...
... ... the ... ...
... the ... ...
... ...
... after ... ...
... ...
hydrate and sodium
carbonate had no action ... ...
... .

... ...
... ...
... ... of ... ...
... dissolved in absolute
alcohol, and treated with ...
... ... of ...
potassium ... ... , the
solution was boiled for five
minutes, and dark ...

into ... a ... of ... te..
... d flocculent precipitate
settled to the bottom of th
essel, leaving the supernatant
liquid clear and recover
... The ...
... was ... d by de... a ...
until no longer alkaline, ...
no... d with water, and wa...
again, after filtering ...
d...ing, its solu bility, in
... of solvents ... s
tected, for t... ... pos... of
... fying it. ... ted ...
... o. ... l, t... ... te... ...
... ... d ... q... ... ...
... o... to ... ...
... s... appe... ... p...
of t... ... t o... ...

D ann. Chem. (Liebig). 100. 208

Pyramidoöxanthotauylp... ... ...

Four grams of pyramidoöx antho-
tauylp... ... were dissolved
in 10 cc. of ... alcohol, about
20 cc. of concentrated ... ...
... added, and sulphuretted
hydrogen passed into the solution
for fifteen minutes. The solution
became brood-red in a ...
while, and increased considerably
in temperature. After the sulphuretted
hydrogen had been passed into the
solution no sulphur was ...
made to separate ... ...,
whereas it came out by the ...
solution ... on a ...

of t... in pe ... . A small
amount of the solution was
removed and treated with
water, which cau... ... ...
precipitate of ... ...
... ... to for... ... e
... ... of the solutio...
was then evaporated to dryness
on a water-bath, and the
residue digested with water
containing a small amount
of hydrochloric acid, ...
po...ion ... ...aining ... ...
... ...sisted of ...,
...as removed by filtration
through filters of ...,
...mai...ng the filtrate
...water with ...

and o ... the ... ...
... and H ... ...,
and the compound ... ... d
... in the ... of ... ...

    Determination of nitrogen
gave the following results, –
0.1681 grams gave 5 c.c. N at 17°C. and
     755 mm. Hg. pressure = ?.?(?.?% ...
?.?+?? grams gave 11.5 c.c. N at ?°
  as d 755 mm. Hg. pressure = ?.???? grams.
  Calculated for   Found.

$C_1 H_{10} {?}_2 N_{?}$	L.	II.
N. ?.?%	?	?

    From its formula,   it
appears that this compound is
– ...rivative of para toluidine.
... as such it ...ad... form...
... ... with ... ...

Action of Oxidizing Agents on Para nitro orthophenyl sulphone

Chromic acid

It was found that the best method for converting paranitro phenylsulphuric... into paraphenyl sulphonevenzoic acid was by treatment with chromic acid in a in acetic acid solution. ... method was found it this is sulphone.

Ten grams of ... were dissolved in g... acetic, and ... at ... to ... in a flask with ... to ... ted ...

ated solution of a
gramme of ... ...
in ... ... a time
added in small
... at a time,
on being ... for
... to boil ...
... the chromic ...
added the solut...
... for the ...
... was ... ...
the acetic acid ...
... from the ...
... precipitate
... to consist of ...
Cr_2O_3, almost ...
organic matter,

<u>Potassium permanganate &c.</u>

The modification ... ... for orthotolidine ... to ..., has generally ... found to be potassium permanganate in water solution.

The action of oxidizing agents on sulphones ... first investigated by ... ? He found that when ... sulphone mixed with water acidified with ... acid was treated with potassium permanganate the permanganate was somewhat ... , as ... by the precipitation of manganese dioxide, ...

and that ... state
that the ... phenyl ...
and tolyl ... ...
oxidized by potassium
permanganate to ...
... required a
phenylsulphonedibenzoic and
respectively, –

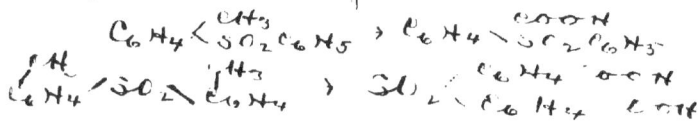

$$C_6H_4 < \genfrac{}{}{0pt}{}{CH_3}{SO_2C_6H_5} \, , \; C_6H_4 < \genfrac{}{}{0pt}{}{COOH}{SO_2C_6H_5}$$

$$C_6H_4 / SO_2 < \genfrac{}{}{0pt}{}{CH_3}{C_6H_4} \, , \; SO_2 < \genfrac{}{}{0pt}{}{C_6H_4 \, COOH}{C_6H_4 \, COOH}$$

When the action was carried
on in acid ... solution the
... was smalled ... water
solution of potassium
permanganate gave ...
... results, although the
... ...
... in water ... ...

...ha t of "I was ex...ing a...
theoretical quantity of
potassium permanganate...

Paranitroorthotolylphenon
sulphone was found to a...in
a similar manner.

Moran ski and itingl[1]
have determined by
experiment that one
mole cue of potassium
permanganate gives...
and a half atoms of its oxy g
to an oxidizable substance
even no free mineral
acid is present. It would
therefore require two molecules
of the permanganate to...
of the sulphone if all of the
oxygen liberated was as in

... ... required. This is to be a ...
for partly by the fact that
the sulphone is probably
completely broken down to a
certain extent into higher
oxidation-products①, and also
that the manganese dioxide
first formed itself ... ...
the potassium per... not ②
somewhat

<u>Sawanite orthophosphyle</u> l
<u>benzoic acid.</u> C₂H₃... ...(o)

The acid was prepared in
the following method:

Ten grams of the sub—
... suspended in about a
litre of water in a ....;
sulphur had previously been
powdered as fine as was
possible by trituration in
a mortar. A solution of
thirty grams of potassium
permanganate was added,
one quarter at a time, at
intervals of an hour, the
solution being rested to
boiling meanwhile o — — —
— ...... The color of the perman
ganate was gradually

dissolved by the ... ...
manganese-dioxide as the
reaction progressed. It was
difficult to determine from
the appearance of the
solution when the ...
traces of the permanganate
had disappeared, without
letting the solution settle, which
required considerable time.
... samples were removed
... occasionally and filtered,
and when a colorless filtrate
was obtained the action was
stopped. It usually
required eight hours to
... complete solution of a ...
permanganate ... ... so that
was allowed to settle ...
...

...ng, and then the ... supernatant liquid... off the ... over water. On acidifying the filtrate with ... acid a white, finely divided precipitate ... , which aggregated into ... quickly ... shaped crystals ... forming. This was filtered off and purified by ... ing in water, which ... left some of the sulphur undissolved, and treating ... with ... charcoal. The largest yield of acid obtained from 16 grams of the sulphur ... was ...8 grams. A considerable amount of sulphur ... as ... water ... ...

... mixed with t... ...
dioxide precipitate, ... was
recovered by dissolving
the dioxide in ... ...
... ... and filtering off the
undissolved solution,
which was not attacked by
the nascent chlorine. Otto[C]
found that phenylsulphone
also is not attacked by
nascent chlorine in solution,
but on the other hand that
dry chlorine acts on the
sulphone at its melting-point
in diffused light to form
benzene-sulphone chloride
and monochlorbenzene—[2]

Paranitroorthophenyl-
sulpho-benzoic acid
crystallizes in white needles
5 to 10 mm. long, melting at
130° mercury. It is difficult
soluble in cold water, and
very readily in hot
water. It has a bitter ... ...
taste. Acetone dissolves it
readily, but ether and
benzene only to a small
extent

<u>Salts</u> of ta... ...

Barium nitroprussi-sulf... ...
...oote, (C₄... ...(... ...) Ba.N...
...as obtain d by d'ss... ...
barium ... ...ate in ...
solution of th a id, ...
crystalli,es in th... form
... V, white ... ...line, ...
together, and diffic... ...
so... in ... water. ... ...
... the following results,

I. 0.3784 grams lost 0.0100 grams at 116°
        and gave 0.114 grams ... ...

II. 0.30 ... grams lost ... ... grams at 11°
        ... gave 0.1... ... grams ...

Calculated for   Found.

$(C_{15}H_8(O_2N)_2)_2 Ba . H_2O$ I   II

H₂O   2.31   ....

$(C_{15}H_8(O_2N)_2)_2 Ba$

Ba.   18...2   18...   1...

___ium nitro ___ ___
benzoate, $(C_{15}H_8(NO_2)_2CH_2 ...)_2 Ba . 1/2 H$.
was obta____ ____ boiling ____
solution of t__ acid ____ th__ __
__x ___ of ____ ___, __ ___ crystallizes
in tufts of radiating, ___ less,
prismatic crystals, ____
___ ___ ___ water.

A ___ of th___ ____ ___
th___ following ____, ___
0.1401 gr ams ___ ___ 0.__ grams at 16°.
___ ___ ___ st __ ___ ___

Comparison of Paranitro-orthophenyl sulphone acid with Paranitro... benzoyl ... sulphon...

Hollis found that th...
formed by the action of...
and dilute acids on th...
product of the Friede...
crafts reaction with th...
...monides of paranitron t...
sulphobenzoic acid, was...
easily soluble in water, ...
could not be obtain...
a crystallized form, its...
...num salt crystallized
with varying amount...
water of crystallisation,
...less than ...

The crystals ...
... of ... so
as ... , to
... crystals ...
... plates with ...
... ties of water of
crystallization.

... benzoic ...
benzoic acid, on the other ...,
is ... difficultly ...
in water, and crystallizes ...
... . Its ...
crystallizes ... ...
... together, containing 9
... of water,
crystallization, ... its
... said ... ...
... taking it
... ... of

... rate of ... ...

The difference ...

~~...~~ ... equal ... more striking ... ... of a salt, ...

<u>Pyro- & ortho-</u> ... ... the ... ... su[l]phuric Acid. ...

... soluble in | difficulty, ... water. | in water.

... crystallizes ... rusts ...

... uranium sa... ...
... tains ... ...
to seven molecules ... ... of ... water or ... of crystal ...
crystallization, ...

_____ ___ _____ __ __

__ _____ __ ___ __ _

_____ _____ __ ___

acid.

Phosphorus pentacle oxide

acts readily on the peroxide

to give _____ __ _

_____ _____ _____.

When the acid was

intimately mixed with

phosphorus _____ _____ __

from _____ _____ __

took _____ _____ If

temperature ____ was raised

16 °C, but when the ____ to

_____ ___tained sharp

explosion _____ the reaction

_____ _____ _____ __ _

__ _____ __ T_ _____ ____ __ __ __

$$C_6H_5 \cdot N \cdots + \cdots \quad ... \quad + ...$$

_____ _____ _____
__ _____ __ chloride.

Equal weights of ___
nitro __ sulphur ____
benzoic acid and ____ __
__ __ _____ _____
_____ _____ of ___
____ __ __ _____ __
___ __ a _____ __
_____ ___ __ ___ ___
___ a ___ ____ __ ___ __
___ __ _____

It crystallised ... of ...
solution on the ... ...
evaporation of the ...
... well-formed, ...,
... crystals ...
... millimetre ... ...,
... melting ...
114°. (...) The crystals were
easily soluble in chloroform,
ether, benzene and ...,

... were determined
... ... by ...
with a dilute solution of
... soda ...
... ...,
acidified ... with nitric
acid ... precipitate ... ...
silver nitrate.

I. 0.1631 grams gave 0.0916 grams CO2

II 0.1877 grams gave 0.0810 grams CO2

Calculated for   Found

$C_{13}H_{15}NO_3Cl$ I.   II

C   10.66   10.85   10.54

Action of Water on Silver chloride cupric chloride.

The chloride is quite stable toward water at ordinary temperatures. No change could be detected in a ... stable ti al ... ... nts ... with water several days. But it is readily decomposed ... ... water.

... ...

solution decompose the
chloride much more ___ ___
than water. Two-tenths of a
gram of the chloride was
___ decomposed and
passed into solution ___
boiled with a very little
solution of caustic potash for
five minutes. Hydrochloric
acid was added to these, etc.
in ___, and very fine
___-shaped crystals
separated out on cooling. ___ ___
melting-point, 110° C, showed
them to be paranitroortho
phenylsulphonebenzoic acid
___ reaction with water ___
represented thus,

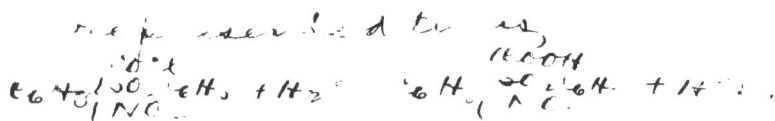

$$C_6 + ? SO_2 \cdot C_6H_5 + H_2 O \quad C_6H_4 \cdot SO_2 \cdot C_6H_4 + H ? .$$
$$| NO_2 \qquad \qquad COOH$$

...tion of ...... on a ...
....
..... **Chlorine**.

By the action of ....
the ...amine of the carbonyl
chloride group is ....
by an amido residue.

Some of the finely ......
...... was added to a ...
solution of a ....
...... and ...... a ....
...... with ....
...... The ...... done ....
...... of the time ...... the ....
appear... as the ....
a light yellow ......,
... . was ....
...... on ...., ... the
solution filtered from the

... ...
... diss... d ... ...
l ... ..., ...
... ...
... ... crystallized ...
on spontaneous evaporation
of ... ... solution ... ...
... ... , ..., ...
... crystals, apparently , mono —
clinic , and melting at
111° $14'$ ... (uncor.) The crystals
were analysed with the ... ...
... results,

I 0.4044 grams gave ... grams $BaSO_4$.
II ... grams gave ... grams ... .

	Calculated for $C_{13}H_{10}C_5N_2O$	Found I	II
O	10.4?	10.14	
N	4.1?		1.14.

The reaction, therefore, which
takes place is,

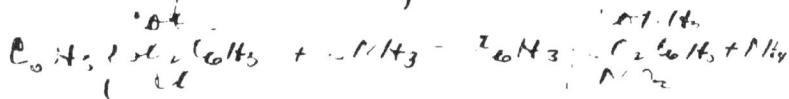

... ...
... ...

... nitrogen ...
... oxide.

... oxide is ... ...
insoluble in water, ...,
... ... It dissolves ... easily
in ... alcohol, ...
... ... as the
solution cools, in th... form a
... dissol... ...
very ... , ...
... ...
... ...

all the ... other ... of the weak ...
was found to me in ...
... experiment ... , they ...,
... it is often ... acids. It
did not therefore ... that
to ... investigation.

... of material
prevented the ... of
of my more experiments
of this kind.

In the ... and ...
reaction every ... substance
... to require different ...
conditions, ...
... mixed experience ...
for each case. And for this
reason the ... of ...
... single experiments
... to consider

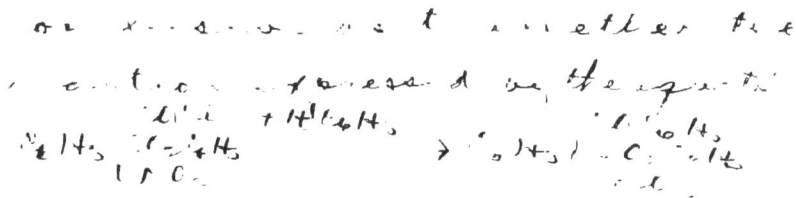

Comparison of ...

... compound obtained
by Hollis[1] from the ...

[1] Lee ...

sulphorous ... ...,
isomeric with it ... ...
compound, crystallizes ...
... ...
... ... ta
... ... at 170°C. It forms a
... crystallizes ... 'd, ...
... with water, ... to
... 140°... ; and ... ... 't
... ... to form ... ...
crystallizing in prisms
... ... at 141° to 14..°...
These properties characterize
it as quite distinct from the
compound of the same
empirical composition
obtained ... ...,

potassium salt of the
para...benzoic...acid... the
chloride, $C_6H_5...COCl$, is formed.
It was hoped that ...
...tion of benzoin...
...aluminum...
would yield, ...
although...employ...
according to the reaction

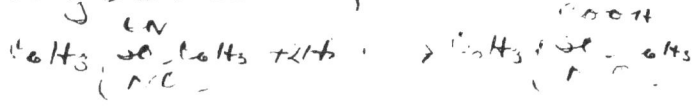

$$C_6H_3(CN)_2 + H_2C_6H_5 \rightarrow C_6H_3 \cdot C_6H_5 \cdot C_6H_5$$

...which by hydrolysis o the
...acid...
...to...benzoic acid,

$$C_6H_3(CN)_2 \cdot C_6H_5 + H_2O \rightarrow C_6H_3(COOH)_2 \cdot C_6H_5$$

The material for the
investigation was prepared
...

t...is ... ... atom, Th...
ammonium salt... ...
was obtained as a ...t_
_broad ... in the ... ratio
of th... symmetrical ...
of ... aminoanthracene ...
... prepared, was ... serted
into the potassium salt ...
...ding ... caustic potassa
potassium ... ...ts
solution in hot water. Th...
potassium salt is difficultly
soluble in cold water, and ...
separates out as th... solution
cools in th... form ...
Potassium chloride was found
to work more satisfactorily
than potassium ... ... The
salt was ... ...

[illegible] absent [illegible]

[illegible]

[illegible]

[illegible]

[illegible] after [illegible]

[illegible] from [illegible]

[illegible] of the [illegible]

[illegible]

[illegible] the Frieden [illegible] to

[illegible]

[illegible] brown [illegible]

[illegible] to [illegible]

[illegible] on the [illegible]

[illegible]

[illegible] ted [illegible]

[illegible]

[illegible] of the [illegible] ed

[illegible]

[illegible]

5. Paranitroorthophenyl sulpho...

...enjoy...

...definite prop...ties

which distinguish...

the isomeric substan...

...ined by the action of...

...and...

...on the...

of paranitroortho...

...d, especially...

...tion of an amide...

...tion of ammonia...

the other forms a...

Biographical.

Robert Stewart Norris... a
born in Belleview, N.C.,
January 19, 1869. He received
his primary education in
the public schools of John ...,
Kansas, and prepared to ...
... The ... and ...
department of Missouri ...
College at the same ...
In 1887 ... removed to Los
Angeles, California, ...
entered the University of
California in 1..., graduated
... in 18.. with the d...
of D.C. He remained ... ...
own institution as assistant
in chemistry for three
years, and entered ...

... ...
... ...
courses in ... history,
... biology, and
... physiology. He was
appointed ... ...
Scholar ... ...
... ...

Baltimore
May 1897